Clean Energy Innovations

Pioneering the Path to a Sustainable Tomorrow

Maia Hoffman

Clean Energy Innovations

TABLE OF CONTENTS

Advances in Photovoltaic Efficiency

The sun, a colossal fusion reactor blazing in the sky, has been a source of energy for millennia. Harnessing its power efficiently has been a quest for scientists and engineers, leading to remarkable advances in photovoltaic technology. The journey to improve photovoltaic efficiency is a tale of innovation, persistence, and the relentless pursuit of harnessing the sun's energy more effectively.

Photovoltaic cells, the building blocks of solar panels, convert sunlight directly into electricity. The efficiency of these cells is a measure of how much sunlight they can convert into usable energy. Over the years, significant strides have been made in enhancing this efficiency, driven by the need for sustainable and renewable energy sources.

One of the most notable advancements in photovoltaic efficiency is the development of multi-junction solar cells. Unlike traditional single-junction cells, which are limited by the Shockley-Queisser limit, multi-junction cells stack multiple layers of semiconductor materials, each designed to capture different segments of the solar spectrum. This layered approach allows for a broader absorption of sunlight, significantly boosting efficiency. Some multi-junction cells have achieved efficiencies exceeding 40%, a remarkable feat compared to the average 15-20% efficiency of conventional silicon-based cells.

The exploration of new materials has also played a crucial role in advancing photovoltaic efficiency. Perovskite solar cells, named after the mineral with a similar crystal structure, have emerged as a promising alternative to traditional silicon cells. These materials are not only cheaper to produce but also exhibit impressive efficiencies, with some laboratory tests reaching over 25%. The flexibility and lightweight nature of perovskite cells open up new possibilities for solar applications, from wearable technology to building-integrated photovoltaics.

Another breakthrough in photovoltaic efficiency is the use of tandem solar cells. By combining different types of solar cells, such as perovskite and silicon, in a single device, tandem cells can surpass the efficiency limits of individual cells. This synergy allows for more effective utilization of the solar spectrum, pushing the boundaries of what is possible in solar energy conversion.

The integration of nanotechnology has further propelled advancements in photovoltaic efficiency. Nanostructures, such as quantum dots and nanowires, have unique optical and electronic properties that can enhance light absorption and charge carrier mobility. These nanoscale innovations enable the design of solar cells with improved performance and reduced material usage, paving the way for more efficient and cost-effective solar solutions.

In addition to material innovations, advancements in photovoltaic efficiency have been driven by improvements in manufacturing processes. Techniques such as thin-film deposition and roll-to-roll printing have revolutionized the production of solar cells, making them more affordable and accessible. These methods allow for the creation of flexible and lightweight solar panels that can be easily integrated into various surfaces, from rooftops to vehicles.

The quest for higher photovoltaic efficiency is not without its challenges. Stability and durability remain critical concerns, particularly for emerging materials like perovskites. Researchers are actively working to address these issues, developing encapsulation techniques and protective coatings to enhance the longevity of solar cells. Additionally, the environmental impact of solar cell production is being scrutinized, with efforts to develop more sustainable and eco-friendly manufacturing processes.

Despite these challenges, the future of photovoltaic efficiency looks promising. The continuous evolution of solar technology is driven by a global commitment to reducing carbon emissions and transitioning to renewable energy sources. As photovoltaic efficiency improves, solar energy becomes an increasingly viable and competitive option, contributing to a more sustainable and resilient energy future.

The impact of these advancements extends beyond the realm of energy production. Improved photovoltaic efficiency has the potential to transform industries, economies, and societies. In regions with limited access to electricity, efficient solar technology can provide a reliable and affordable source of power, fostering economic development and improving quality of life. In urban environments, the integration of high-efficiency solar panels into buildings and infrastructure can reduce energy consumption and carbon footprints, contributing to the creation of smart and sustainable cities.

The journey towards higher photovoltaic efficiency is a testament to human ingenuity and the relentless pursuit of progress. As researchers and engineers continue to push the boundaries of what is possible, the potential for solar energy to power our world becomes increasingly tangible. The sun, once a distant and untouchable force, is now within our grasp, offering a clean and abundant source of energy for generations to come.

In the grand tapestry of renewable energy, advances in photovoltaic efficiency are a shining thread, weaving together innovation, sustainability, and hope. The sun's energy, harnessed with ever-increasing efficiency, holds the promise of a brighter and more sustainable future, illuminating the path towards a world powered by clean and renewable energy.

Emerging Solar Materials and Designs

The quest for harnessing solar energy has led to a fascinating exploration of emerging materials and innovative designs. As the demand for renewable energy sources grows, researchers and engineers are delving into the potential of new materials that promise to revolutionize the solar industry. These advancements are not only enhancing the efficiency of solar panels but also expanding their applications, making solar energy more accessible and versatile.

One of the most promising materials in the solar landscape is perovskite. Named after the mineral with a similar crystal structure, perovskite materials have garnered significant attention due to their remarkable light absorption properties and ease of fabrication. Unlike traditional silicon-based solar cells, perovskite cells can be manufactured using low-cost, solution-based processes, which significantly reduces production costs. This affordability, combined with their impressive efficiency, makes perovskite solar cells a game-changer in the quest for sustainable energy solutions.

The versatility of perovskite materials extends beyond cost and efficiency. Their flexibility and lightweight nature open up a myriad of possibilities for solar applications. Imagine solar panels that can be seamlessly integrated into the fabric of buildings, transforming windows and facades into energy-generating surfaces. This concept, known as building-integrated photovoltaics (BIPV), is gaining traction as architects and designers seek to create energy-efficient structures that harmonize with their surroundings. The aesthetic appeal of perovskite-based BIPV systems, coupled with their energy-generating capabilities, offers a glimpse into the future of sustainable architecture.

Another exciting development in solar materials is the use of organic photovoltaics (OPVs). These solar cells are composed of carbon-based materials, which offer unique advantages over traditional inorganic cells. OPVs are lightweight, flexible, and can be produced using roll-to-roll printing techniques, similar to how newspapers are printed. This manufacturing process allows for the creation of large-area solar panels at a fraction of the cost of conventional methods. While OPVs currently have lower efficiencies compared to silicon and perovskite cells, ongoing research is focused on enhancing their performance and stability, paving the way for their widespread adoption.

The exploration of emerging solar materials is not limited to perovskites and organics. Quantum dots, nanoscale semiconductor particles, are also making waves in the solar industry. These tiny particles have unique optical properties that can be tuned by simply changing their size. Quantum dot solar cells have the potential to achieve high efficiencies by capturing a broader range of the solar spectrum. Additionally, their ability to be incorporated into flexible and transparent films opens up new possibilities for solar applications, from portable electronics to smart windows.

Innovative designs are complementing the development of new materials, further pushing the boundaries of solar technology. One such design is the bifacial solar panel, which captures sunlight from both sides of the panel. By utilizing reflected light from the ground or surrounding surfaces, bifacial panels can generate more electricity compared to traditional single-sided panels. This design is particularly effective in environments with high albedo, such as snowy or sandy regions, where the ground reflects a significant amount of sunlight.

Concentrated solar power (CSP) systems represent another innovative approach to solar energy generation. Unlike photovoltaic systems that convert sunlight directly into electricity, CSP systems use mirrors or lenses to concentrate sunlight onto a small area, generating heat that is then used to produce electricity. This technology is particularly suited for large-scale power generation and can be integrated with thermal storage systems to provide electricity even when the sun is not shining. The ability to store energy in the form of heat offers a significant advantage in terms of grid stability and reliability.

The integration of solar technology into everyday objects is also gaining momentum. From solar-powered backpacks to solar-charging stations for electric vehicles, the possibilities are endless. These innovations not only provide convenient and sustainable energy solutions but also raise awareness about the potential of solar power in our daily lives. As solar technology becomes more ubiquitous, it has the potential to transform the way we think about energy consumption and production.

The journey towards harnessing the full potential of solar energy is a testament to human ingenuity and the relentless pursuit of progress. Emerging solar materials and designs are at the forefront of this journey, offering new opportunities for sustainable energy generation. As researchers and engineers continue to push the boundaries of what is possible, the future of solar energy looks brighter than ever.

The impact of these advancements extends beyond the realm of energy production. The development of new solar materials and designs has the potential to transform industries, economies, and societies. In regions with limited access to electricity, affordable and efficient solar technology can provide a reliable source of power, fostering economic development and improving quality of life. In urban environments, the integration of solar panels into buildings and infrastructure can reduce energy consumption and carbon footprints, contributing to the creation of smart and sustainable cities.

The exploration of emerging solar materials and designs is a testament to the power of innovation and the potential of renewable energy. As we continue to harness the sun's energy with ever-increasing efficiency and creativity, we move closer to a future powered by clean and sustainable energy sources. The sun, once a distant and untouchable force, is now within our grasp, offering a clean and abundant source of energy for generations to come.

Innovations in Solar Storage Solutions

Solar energy, with its promise of clean and abundant power, faces a significant challenge: the sun doesn't shine all the time.

This intermittency has driven the need for effective solar storage solutions, which are crucial for ensuring a steady and reliable energy supply. Innovations in solar storage are transforming the landscape of renewable energy, making it possible to harness the sun's power even when it's not shining.

One of the most significant advancements in solar storage is the development of advanced battery technologies. Lithium-ion batteries, widely used in consumer electronics, have become a cornerstone of solar storage systems. Their high energy density, long cycle life, and decreasing costs make them an attractive option for storing solar energy. However, the quest for even better performance has led to the exploration of alternative battery chemistries.

Sodium-ion batteries are emerging as a promising alternative to lithium-ion technology. Sodium is more abundant and less expensive than lithium, making sodium-ion batteries a cost-effective option for large-scale energy storage. While they currently have lower energy densities compared to their lithium counterparts, ongoing research is focused on improving their performance and scalability. The potential for sodium-ion batteries to provide affordable and sustainable energy storage solutions is driving significant interest and investment.

Flow batteries represent another innovative approach to solar storage. Unlike traditional batteries, which store energy in solid electrodes, flow batteries use liquid electrolytes stored in external tanks. This design allows for easy scalability by simply increasing the size of the tanks, making flow batteries ideal for large-scale solar installations. Vanadium redox flow batteries, in particular, have gained attention for their long cycle life and ability to maintain capacity over time. Their flexibility and durability make them a compelling option for integrating solar energy into the grid.

Thermal energy storage is also playing a crucial role in solar storage solutions. Concentrated solar power (CSP) systems, which use mirrors or lenses to focus sunlight and generate heat, can store this thermal energy for later use. Molten salt is a popular medium for thermal storage in CSP systems, as it can retain heat for extended periods. This stored heat can then be used to produce electricity even when the sun is not shining, providing a reliable and continuous power supply. The ability to store energy in the form of heat offers a significant advantage in terms of grid stability and reliability.

The integration of solar storage solutions with smart grid technology is further enhancing the efficiency and reliability of solar energy systems. Smart grids use advanced communication and control technologies to optimize the distribution and consumption of electricity. By integrating solar storage with smart grids, energy providers can better manage supply and demand, reduce peak loads, and improve grid resilience. This synergy between solar storage and smart grids is paving the way for a more flexible and adaptive energy infrastructure.

Innovations in solar storage are not limited to large-scale applications. Residential solar storage systems are becoming increasingly popular as homeowners seek to maximize their energy independence and reduce reliance on the grid. Home battery systems, such as Tesla's Powerwall, allow homeowners to store excess solar energy generated during the day for use at night or during power outages. These systems provide a sense of security and self-sufficiency, empowering individuals to take control of their energy consumption.

The rise of electric vehicles (EVs) is also influencing the development of solar storage solutions. As EV adoption grows, the demand for efficient and sustainable charging infrastructure is increasing. Solar-powered charging stations, equipped with storage systems, offer a clean and renewable solution for powering electric vehicles. These stations can store solar energy during the day and provide electricity for charging vehicles at any time, reducing the strain on the grid and promoting the use of renewable energy in transportation.

The potential of solar storage solutions extends beyond energy independence and grid stability. By enabling the integration of renewable energy sources, solar storage is playing a vital role in reducing carbon emissions and combating climate change. As countries around the world strive to meet their climate goals, the development and deployment of innovative solar storage technologies are becoming increasingly important.

The journey towards effective solar storage solutions is a testament to human ingenuity and the relentless pursuit of progress. As researchers and engineers continue to push the boundaries of what is possible, the future of solar energy looks brighter than ever. The ability to store and utilize solar energy efficiently is transforming the way we think about energy production and consumption, paving the way for a more sustainable and resilient energy future.

The impact of these innovations extends beyond the realm of energy production. The development of advanced solar storage solutions has the potential to transform industries, economies, and societies. In regions with limited access to electricity, affordable and efficient solar storage can provide a reliable source of power, fostering economic development and improving quality of life. In urban environments, the integration of solar storage systems into buildings and infrastructure can reduce energy consumption and carbon footprints, contributing to the creation of smart and sustainable cities.

The exploration of solar storage solutions is a testament to the power of innovation and the potential of renewable energy. As we continue to harness the sun's energy with ever-increasing efficiency and creativity, we move closer to a future powered by clean and sustainable energy sources. The sun, once a distant and untouchable force, is now within our grasp, offering a clean and abundant source of energy for generations to come.

The Rise of Solar Microgrids

In the quest for sustainable energy solutions, solar microgrids have emerged as a transformative force, reshaping how communities generate and consume electricity. These localized energy systems, powered primarily by solar panels, offer a decentralized approach to energy distribution, providing resilience, reliability, and independence from traditional power grids. The rise of solar microgrids is a testament to the growing demand for clean energy and the innovative ways in which it can be harnessed.

At the heart of a solar microgrid lies the concept of localized energy generation. Unlike traditional power grids that rely on centralized power plants and extensive transmission networks, microgrids generate electricity close to where it is consumed. This proximity reduces transmission losses and enhances the efficiency of energy delivery. Solar panels, with their ability to convert sunlight directly into electricity, are ideally suited for microgrid applications, offering a renewable and abundant source of power.

One of the most compelling advantages of solar microgrids is their ability to operate independently of the main power grid. This feature, known as islanding, allows microgrids to continue supplying electricity even during grid outages. In regions prone to natural disasters or grid instability, this capability is invaluable, providing a reliable source of power when it is needed most. Communities equipped with solar microgrids can maintain essential services, such as hospitals, emergency response centers, and communication networks, ensuring safety and continuity during crises.

The flexibility of solar microgrids extends beyond their ability to operate independently. These systems can be tailored to meet the specific energy needs of a community, whether it be a remote village, an industrial complex, or a university campus. By integrating solar panels with energy storage solutions, such as batteries, microgrids can store excess energy generated during sunny periods for use during cloudy days or at night. This capability not only enhances energy reliability but also maximizes the utilization of solar power, reducing reliance on fossil fuels and lowering carbon emissions.

The rise of solar microgrids is also driven by the increasing affordability of solar technology. Over the past decade, the cost of solar panels has plummeted, making them an economically viable option for a wide range of applications. This cost reduction, coupled with advancements in energy storage and grid management technologies, has made solar microgrids an attractive solution for both developed and developing regions. In areas with limited access to electricity, microgrids offer a pathway to energy independence, fostering economic development and improving quality of life.

The implementation of solar microgrids is not without its challenges. Regulatory hurdles, financing constraints, and technical complexities can pose significant barriers to deployment. However, innovative business models and supportive policy frameworks are emerging to address these challenges. Public-private partnerships, community ownership models, and performance-based incentives are helping to overcome financial and regulatory obstacles, paving the way for widespread adoption of solar microgrids.

The impact of solar microgrids extends beyond energy generation. By empowering communities to take control of their energy resources, microgrids promote local economic development and job creation. The installation, operation, and maintenance of microgrid systems require skilled labor, providing employment opportunities and fostering the growth of local industries. Additionally, the decentralization of energy generation can reduce energy costs for consumers, freeing up resources for other essential needs.

The environmental benefits of solar microgrids are equally significant. By reducing reliance on fossil fuels, microgrids contribute to a reduction in greenhouse gas emissions and air pollution. This shift towards cleaner energy sources is crucial in the fight against climate change and the pursuit of a sustainable future. Furthermore, the integration of solar microgrids with other renewable energy sources, such as wind or biomass, can enhance the overall sustainability and resilience of energy systems.

The rise of solar microgrids is a testament to the power of innovation and the potential of renewable energy. As communities around the world embrace this technology, they are not only transforming their energy landscapes but also setting an example for others to follow. The success of solar microgrids demonstrates that a sustainable and resilient energy future is within reach, driven by the collective efforts of individuals, businesses, and governments.

The journey towards widespread adoption of solar microgrids is a dynamic and evolving process. As technology continues to advance and costs continue to decline, the potential for microgrids to revolutionize energy systems becomes increasingly tangible. The lessons learned from early adopters and pilot projects are informing best practices and guiding future developments, ensuring that solar microgrids remain at the forefront of the clean energy transition.

In conclusion, the rise of solar microgrids represents a paradigm shift in how we think about energy generation and consumption. By harnessing the power of the sun and embracing decentralized energy solutions, communities are taking charge of their energy futures, building resilience, and contributing to a more sustainable world. The potential of solar microgrids is vast, and as we continue to innovate and adapt, the possibilities for a cleaner, greener energy future are limitless.

Case Studies: Cutting-Edge Solar Projects

In the ever-evolving landscape of renewable energy, certain solar projects stand out as beacons of innovation and progress.

These cutting-edge initiatives not only push the boundaries of technology but also demonstrate the transformative potential of solar energy in diverse contexts. By examining these case studies, we gain valuable insights into the creative solutions and strategic approaches that are shaping the future of solar power.

One remarkable example is the Noor Abu Dhabi solar plant in the United Arab Emirates. As the largest single-site solar project in the world, Noor Abu Dhabi spans over 8 square kilometers and boasts a capacity of 1.17 gigawatts. This massive installation utilizes over 3.2 million solar panels, harnessing the abundant sunlight of the desert to power approximately 90,000 homes. The project's success lies in its scale and efficiency, achieved through the use of advanced photovoltaic technology and innovative design. By optimizing panel placement and incorporating cutting-edge tracking systems, Noor Abu Dhabi maximizes energy capture, setting a new standard for large-scale solar installations.

In a different part of the world, the Solar Settlement in Freiburg, Germany, offers a compelling example of how solar energy can be integrated into urban living. This pioneering community consists of 59 homes, each equipped with solar panels that generate more energy than the residents consume. The surplus electricity is fed back into the grid, creating a positive energy balance. The Solar Settlement exemplifies the concept of a "plus-energy" community, where buildings are designed not only to be energy-efficient but also to produce excess renewable energy. This project highlights the potential for solar power to transform residential areas into sustainable, self-sufficient communities.

The floating solar farm on the Queen Elizabeth II Reservoir in the United Kingdom presents another innovative approach to solar energy generation. Covering an area equivalent to eight football fields, this floating array consists of over 23,000 solar panels. The project addresses the challenge of land scarcity by utilizing water surfaces for solar installations. Floating solar farms offer several advantages, including reduced land use, minimized environmental impact, and enhanced panel efficiency due to the cooling effect of water. This case study demonstrates the versatility of solar technology and its ability to adapt to different environmental constraints.

In the realm of solar transportation, the Solar Impulse project has captured global attention. This groundbreaking initiative involved the development of a solar-powered aircraft capable of flying around the world without a single drop of fuel. The Solar Impulse 2, equipped with over 17,000 solar cells, successfully completed its historic journey in 2016, proving the viability of solar energy in aviation. The project's success was not only a technological triumph but also a powerful statement about the potential of renewable energy to revolutionize transportation. By pushing the limits of what is possible, Solar Impulse has inspired further research and development in solar-powered mobility.

The Tesla Gigafactory in Nevada, USA, represents a different facet of solar innovation, focusing on the integration of solar energy with energy storage solutions. This massive facility, powered by a combination of solar panels and wind turbines, produces lithium-ion batteries for electric vehicles and renewable energy storage systems. The Gigafactory exemplifies the synergy between solar power and energy storage, highlighting the importance of integrated solutions in achieving a sustainable energy future. By producing batteries on a large scale, Tesla is driving down costs and making renewable energy more accessible to consumers worldwide.

In India, the Cochin International Airport has set a precedent by becoming the world's first fully solar-powered airport. With a solar plant consisting of over 46,000 panels, the airport generates more electricity than it consumes, feeding the surplus back into the grid. This achievement underscores the potential for solar energy to power large-scale infrastructure projects, reducing carbon footprints and promoting sustainability. Cochin's success has inspired other airports and transportation hubs to explore similar initiatives, paving the way for a greener aviation industry.

The Solar Roadways project in the United States offers a glimpse into the future of infrastructure powered by solar energy. This innovative concept involves the development of solar panels that can be embedded into roads, parking lots, and sidewalks. These panels are designed to withstand the weight of vehicles while generating electricity and providing additional functionalities, such as LED lighting and heating elements to melt snow. While still in the experimental stage, Solar Roadways represents a bold vision for integrating renewable energy into everyday infrastructure, transforming roads into energy-generating assets.

In Africa, the Lake Turkana Wind Power project in Kenya demonstrates the potential for hybrid renewable energy systems. While primarily a wind power project, it incorporates solar energy to enhance its overall capacity and reliability. By combining wind and solar resources, the project ensures a more consistent and stable energy supply, addressing the intermittency challenges associated with individual renewable sources. This case study highlights the benefits of hybrid systems in optimizing renewable energy generation and meeting the energy needs of diverse regions.

These case studies illustrate the diverse and innovative ways in which solar energy is being harnessed around the world. From large-scale power plants to community-based initiatives and cutting-edge transportation solutions, solar projects are redefining the possibilities of renewable energy. Each project offers unique insights into the challenges and opportunities associated with solar technology, providing valuable lessons for future developments.

The success of these initiatives is a testament to the power of collaboration and innovation. By bringing together governments, businesses, researchers, and communities, these projects have overcome technical, financial, and regulatory hurdles to achieve remarkable outcomes. As the global demand for clean energy continues to grow, the lessons learned from these case studies will inform and inspire the next generation of solar projects, driving the transition towards a sustainable energy future.

The journey towards a solar-powered world is an ongoing process, marked by continuous advancements and breakthroughs. As technology evolves and new opportunities emerge, the potential for solar energy to transform our energy systems becomes increasingly tangible. By learning from the successes and challenges of cutting-edge solar projects, we can chart a path towards a cleaner, more sustainable future, powered by the limitless energy of the sun.

Technological Innovations in Wind Turbines

The evolution of wind turbines has been marked by a series of technological innovations that have significantly enhanced their efficiency, reliability, and adaptability. As the world increasingly turns to renewable energy sources, wind power stands out as a key player in the transition to a sustainable energy future. The advancements in wind turbine technology are not only making wind energy more competitive but also expanding its potential applications across diverse environments.

One of the most notable innovations in wind turbine technology is the development of larger and more efficient turbine blades. The size of turbine blades has grown substantially over the years, with some of the latest models boasting blades that exceed 100 meters in length. These longer blades capture more wind energy, increasing the overall power output of the turbine. The use of advanced materials, such as carbon fiber composites, has been instrumental in this development. These materials offer a high strength-to-weight ratio, allowing for the construction of longer blades without compromising structural integrity. The result is a new generation of wind turbines that can generate more electricity with fewer units, reducing the cost of wind energy.

The design of wind turbine blades has also seen significant improvements. Engineers are employing sophisticated aerodynamic modeling techniques to optimize blade shapes, minimizing drag and maximizing lift. This optimization enhances the efficiency of energy capture, even in low-wind conditions.

Additionally, innovations such as serrated trailing edges and vortex generators are being incorporated into blade designs to reduce noise and improve performance. These advancements are making wind turbines more suitable for deployment in a wider range of locations, including areas with lower wind speeds or proximity to residential communities.

Another groundbreaking development in wind turbine technology is the advent of floating wind turbines. Traditional wind turbines are typically installed on fixed foundations, limiting their deployment to shallow waters. Floating wind turbines, however, are anchored to the seabed using mooring lines, allowing them to be placed in deeper waters where wind resources are often more abundant. This innovation opens up vast new areas for wind energy development, particularly in regions with deep coastal waters. Floating wind farms have the potential to significantly increase the global capacity for wind energy, contributing to a more diverse and resilient energy mix.

The integration of digital technologies is also transforming the wind energy sector. Advanced sensors and data analytics are being used to monitor turbine performance in real-time, enabling predictive maintenance and reducing downtime. By analyzing data on factors such as wind speed, temperature, and vibration, operators can identify potential issues before they lead to costly failures. This proactive approach to maintenance not only extends the lifespan of wind turbines but also enhances their reliability and efficiency. Furthermore, the use of artificial intelligence and machine learning algorithms is optimizing turbine operation, adjusting parameters to maximize energy output based on current conditions.

Energy storage solutions are playing an increasingly important role in the integration of wind energy into the grid. The intermittent nature of wind power has long been a challenge, but advancements in battery technology are providing effective solutions. By storing excess energy generated during periods of high wind, these systems ensure a steady supply of electricity even when the wind is not blowing. This capability is crucial for maintaining grid stability and reliability, particularly as the share of wind energy in the overall energy mix continues to grow. The combination of wind power and energy storage is paving the way for a more flexible and resilient energy system.

The development of hybrid renewable energy systems is another exciting trend in the wind energy sector. By combining wind power with other renewable sources, such as solar or hydropower, these systems can provide a more consistent and reliable energy supply. Hybrid systems take advantage of the complementary nature of different energy sources, balancing fluctuations in wind and solar generation to meet demand more effectively. This approach not only enhances the overall efficiency of renewable energy systems but also reduces the need for backup power from fossil fuels, contributing to a reduction in carbon emissions.

The potential of wind energy is not limited to large-scale installations. Small wind turbines are gaining popularity as a decentralized energy solution for homes, farms, and businesses. These compact turbines can be installed on rooftops or in open spaces, providing a clean and renewable source of electricity for individual users. Advances in small wind turbine technology are making these systems more efficient and affordable, empowering individuals and communities to take control of their energy consumption and reduce their carbon footprint.

The global expansion of wind energy is also being supported by innovative financing and policy mechanisms. Governments and financial institutions are recognizing the importance of wind power in achieving climate goals and are implementing measures to encourage investment in the sector. Feed-in tariffs, tax incentives, and renewable energy certificates are among the tools being used to promote wind energy development. These initiatives are helping to level the playing field for wind power, making it a more attractive option for investors and developers.

The journey of technological innovation in wind turbines is a testament to human ingenuity and the relentless pursuit of progress. As engineers and researchers continue to push the boundaries of what is possible, the future of wind energy looks increasingly promising. The advancements in turbine design, digital integration, and hybrid systems are not only enhancing the efficiency and reliability of wind power but also expanding its potential applications across diverse environments.

The impact of these innovations extends beyond the realm of energy generation. By reducing reliance on fossil fuels, wind energy is playing a crucial role in mitigating climate change and promoting environmental sustainability. The development of wind power is also driving economic growth and job creation, providing opportunities for skilled labor in manufacturing, installation, and maintenance. As the global demand for clean energy continues to rise, the wind energy sector is poised to play a leading role in the transition to a sustainable energy future.

The exploration of technological innovations in wind turbines is a testament to the power of innovation and the potential of renewable energy. As we continue to harness the power of the wind with ever-increasing efficiency and creativity, we move closer to a future powered by clean and sustainable energy sources. The wind, once a distant and untouchable force, is now within our grasp, offering a clean and abundant source of energy for generations to come.

Offshore Wind: Harnessing Ocean Breezes

Offshore wind energy has emerged as a formidable force in the renewable energy sector, capitalizing on the vast and untapped potential of ocean breezes. As the world seeks sustainable solutions to meet its growing energy demands, offshore wind offers a promising avenue for clean and abundant power generation. The journey to harnessing ocean breezes is marked by technological advancements, strategic planning, and a commitment to overcoming the unique challenges posed by the marine environment.

The allure of offshore wind lies in the consistent and powerful winds that sweep across open waters. Unlike onshore wind farms, which can be limited by geographical and environmental constraints, offshore installations benefit from higher wind speeds and less turbulence. This results in a more reliable and efficient energy output, making offshore wind a highly attractive option for large-scale power generation. The potential for offshore wind energy is immense, with estimates suggesting that it could meet global electricity demand several times over.

The development of offshore wind farms involves a complex interplay of engineering, environmental, and logistical considerations. One of the most significant challenges is the design and construction of wind turbines capable of withstanding the harsh marine environment. Offshore turbines are typically larger and more robust than their onshore counterparts, with some models featuring blades that span over 200 meters in diameter. These colossal structures are engineered to endure the corrosive effects of saltwater, high winds, and powerful waves, ensuring long-term reliability and performance.

The installation of offshore wind turbines requires specialized vessels and equipment, as well as meticulous planning and coordination. The process begins with the selection of suitable sites, which involves assessing factors such as wind speed, water depth, seabed conditions, and proximity to existing infrastructure. Once a site is chosen, foundations are installed to anchor the turbines securely to the seabed. These foundations can take various forms, including monopiles, jackets, and floating platforms, each suited to different water depths and seabed conditions.

Floating wind turbines represent a groundbreaking innovation in offshore wind technology, enabling the deployment of wind farms in deeper waters where traditional fixed-bottom turbines are not feasible. These floating structures are anchored to the seabed using mooring lines, allowing them to harness the stronger and more consistent winds found further offshore. The development of floating wind technology is expanding the geographical scope of offshore wind energy, opening up new opportunities for countries with deep coastal waters.

The integration of offshore wind energy into the grid presents its own set of challenges and opportunities. Subsea cables are used to transmit electricity from offshore wind farms to onshore substations, where it is then distributed to consumers.

The design and installation of these cables require careful consideration of environmental impacts, as well as technical and logistical factors. Advances in cable technology and grid management are enhancing the efficiency and reliability of offshore wind energy transmission, ensuring a stable and consistent power supply.

Environmental considerations play a crucial role in the development of offshore wind projects. The potential impacts on marine ecosystems, wildlife, and local communities must be carefully assessed and mitigated. Environmental impact assessments are conducted to evaluate the potential effects of wind farm construction and operation, and measures are implemented to minimize disruption to marine life and habitats. The use of innovative technologies, such as acoustic deterrents and habitat restoration, is helping to balance the need for renewable energy with the protection of marine environments.

The economic benefits of offshore wind energy are substantial, contributing to job creation, economic growth, and energy security. The construction, operation, and maintenance of offshore wind farms require a skilled workforce, providing employment opportunities in engineering, manufacturing, and maritime industries. Additionally, the development of a robust offshore wind sector can reduce reliance on imported fossil fuels, enhancing energy independence and resilience.

The global expansion of offshore wind energy is being driven by supportive policy frameworks and ambitious climate targets. Governments around the world are recognizing the potential of offshore wind to contribute to a low-carbon energy future and are implementing measures to encourage investment and development. These include financial incentives, streamlined permitting processes, and international collaboration on research and innovation. The commitment to offshore wind is reflected in the growing number of projects being planned and constructed, with capacity expected to increase significantly in the coming decades.

The journey to harnessing ocean breezes is a testament to human ingenuity and the relentless pursuit of progress. As technology continues to advance and costs continue to decline, the potential for offshore wind energy to transform our energy systems becomes increasingly tangible. The lessons learned from early adopters and pilot projects are informing best practices and guiding future developments, ensuring that offshore wind remains at the forefront of the clean energy transition.

The exploration of offshore wind energy is a testament to the power of innovation and the potential of renewable energy. As we continue to harness the power of the wind with ever-increasing efficiency and creativity, we move closer to a future powered by clean and sustainable energy sources. The ocean breezes, once a distant and untouchable force, are now within our grasp, offering a clean and abundant source of energy for generations to come.

Integrating Wind Power with Smart Grids

The integration of wind power with smart grids represents a pivotal advancement in the quest for a sustainable energy future. As wind energy continues to grow as a significant component of the global energy mix, the need for efficient and reliable grid integration becomes increasingly important. Smart grids, with their advanced communication and automation capabilities, offer a solution to the challenges posed by the variable nature of wind power, enabling a more resilient and adaptable energy system.

At the core of smart grid technology is the ability to monitor and manage energy flows in real-time. This capability is crucial for integrating wind power, which is inherently variable due to fluctuations in wind speed and direction. By utilizing sensors, data analytics, and automated control systems, smart grids can dynamically adjust to changes in wind energy production, ensuring a stable and consistent power supply. This real-time adaptability minimizes the risk of grid instability and enhances the overall reliability of the energy system.

One of the key features of smart grids is their ability to facilitate demand response, a strategy that aligns energy consumption with availability. In the context of wind power, demand response programs can incentivize consumers to adjust their energy usage based on wind energy production levels. For example, during periods of high wind generation, consumers might be encouraged to increase their electricity usage, such as running appliances or charging electric vehicles. Conversely, during low wind periods, demand can be reduced to match the available supply. This flexibility not only optimizes the use of wind energy but also reduces the need for backup power from fossil fuel sources.

Energy storage solutions are another critical component of integrating wind power with smart grids. By storing excess wind energy generated during peak production periods, these systems ensure a steady supply of electricity even when the wind is not blowing. Advances in battery technology, such as lithium-ion and flow batteries, are making energy storage more efficient and cost-effective. Smart grids can intelligently manage these storage systems, discharging stored energy when needed and recharging during periods of surplus generation. This capability enhances grid stability and maximizes the utilization of renewable energy resources.

The integration of wind power with smart grids also involves the development of advanced forecasting and predictive analytics. Accurate wind forecasts are essential for grid operators to plan and manage energy flows effectively. By leveraging machine learning algorithms and meteorological data, smart grids can predict wind energy production with greater precision, allowing for more informed decision-making. These forecasts enable grid operators to anticipate changes in wind generation and adjust grid operations accordingly, reducing the risk of imbalances and ensuring a reliable power supply.

The deployment of smart grid technology requires significant investment in infrastructure and technology. Upgrading existing grid systems to incorporate smart capabilities involves the installation of advanced metering infrastructure, communication networks, and control systems. While the initial costs can be substantial, the long-term benefits of improved efficiency, reliability, and sustainability make it a worthwhile investment. Governments and utilities are increasingly recognizing the value of smart grids and are implementing policies and incentives to support their development and deployment.

The integration of wind power with smart grids also presents opportunities for greater consumer engagement and empowerment. Smart meters and home energy management systems provide consumers with real-time information about their energy usage, enabling them to make more informed decisions about their consumption patterns. By participating in demand response programs and utilizing smart appliances, consumers can actively contribute to grid stability and reduce their energy costs. This increased engagement fosters a more sustainable energy culture and encourages the adoption of renewable energy technologies.

The environmental benefits of integrating wind power with smart grids are significant. By optimizing the use of wind energy and reducing reliance on fossil fuels, smart grids contribute to a reduction in greenhouse gas emissions and air pollution. This shift towards cleaner energy sources is crucial in the fight against climate change and the pursuit of a sustainable future. Additionally, the enhanced efficiency and reliability of smart grids reduce the need for new power plants and transmission infrastructure, minimizing the environmental impact of energy development.

The global transition to smart grids is being driven by a combination of technological innovation, policy support, and market demand. Countries around the world are investing in smart grid projects and initiatives, recognizing their potential to transform energy systems and achieve climate goals. International collaboration and knowledge sharing are playing a crucial role in advancing smart grid technology and overcoming the challenges associated with its deployment.

The integration of wind power with smart grids is a testament to the power of innovation and the potential of renewable energy. As we continue to harness the power of the wind with ever-increasing efficiency and creativity, we move closer to a future powered by clean and sustainable energy sources. The combination of wind power and smart grid technology offers a pathway to a more resilient, adaptable, and sustainable energy system, capable of meeting the demands of a rapidly changing world.

The journey towards integrating wind power with smart grids is an ongoing process, marked by continuous advancements and breakthroughs. As technology evolves and new opportunities emerge, the potential for smart grids to revolutionize energy systems becomes increasingly tangible. By learning from the successes and challenges of early adopters, we can chart a path towards a cleaner, more sustainable future, powered by the limitless energy of the wind.

Overcoming Challenges in Wind Energy Deployment

The deployment of wind energy, while promising and essential for a sustainable future, is fraught with challenges that require innovative solutions and strategic planning. As the world increasingly turns to renewable energy sources, understanding and overcoming these obstacles is crucial for the successful integration of wind power into the global energy mix. From technical and environmental considerations to social and economic factors, the path to widespread wind energy adoption is complex and multifaceted.

One of the primary technical challenges in wind energy deployment is the variability of wind resources. Wind speeds can fluctuate significantly over time and across different locations, affecting the consistency and reliability of energy production. To address this issue, developers are employing advanced meteorological models and data analytics to identify optimal sites for wind farms. By selecting locations with consistent and strong wind patterns, the efficiency and output of wind energy projects can be maximized. Additionally, the development of larger and more efficient turbines is helping to capture more energy from available wind resources, mitigating the impact of variability.

The integration of wind energy into existing power grids presents another significant challenge. Traditional grids were designed for centralized power generation and may struggle to accommodate the decentralized and intermittent nature of wind power. Upgrading grid infrastructure to incorporate smart grid technologies is essential for managing the flow of wind energy and ensuring grid stability. Smart grids enable real-time monitoring and control of energy flows, allowing for dynamic adjustments based on wind energy production levels. This adaptability is crucial for maintaining a reliable power supply and minimizing the risk of grid instability.

Environmental considerations are also a critical aspect of wind energy deployment. While wind power is a clean and renewable energy source, the construction and operation of wind farms can have impacts on local ecosystems and wildlife. Birds and bats, in particular, are at risk of collision with turbine blades. To mitigate these impacts, developers are implementing measures such as careful site selection, turbine design modifications, and the use of technology to detect and deter wildlife. Environmental impact assessments are conducted to evaluate potential effects and ensure that wind energy projects are developed in an environmentally responsible manner.

Social acceptance and community engagement are vital for the successful deployment of wind energy projects. Local communities may have concerns about the visual impact of wind turbines, noise, and potential effects on property values. To address these concerns, developers are engaging with communities early in the planning process, providing information and opportunities for input. Community benefits, such as local job creation and revenue sharing, can also enhance social acceptance and support for wind energy projects. Transparent communication and collaboration with stakeholders are key to building trust and fostering positive relationships.

Economic factors play a significant role in wind energy deployment. The initial costs of wind farm development, including site acquisition, turbine installation, and grid connection, can be substantial. However, the long-term benefits of wind energy, such as low operating costs and energy independence, make it an attractive investment. Governments and financial institutions are increasingly recognizing the value of wind power and are implementing policies and incentives to support its development. These include tax credits, feed-in tariffs, and renewable energy certificates, which help to level the playing field for wind energy and encourage investment.

The rapid pace of technological innovation is driving down the costs of wind energy and making it more competitive with traditional energy sources. Advances in turbine design, materials, and manufacturing processes are enhancing efficiency and reducing production costs. Additionally, the development of energy storage solutions is addressing the challenge of intermittency, ensuring a steady supply of electricity even when the wind is not blowing. As technology continues to evolve, the economic viability of wind energy is expected to improve further, paving the way for increased deployment.

The global expansion of wind energy is also being supported by international collaboration and knowledge sharing. Countries around the world are working together to develop best practices, share research findings, and overcome common challenges. This collaborative approach is accelerating the deployment of wind energy and contributing to the achievement of global climate goals. By learning from the experiences of early adopters and leveraging collective expertise, the wind energy sector is poised for continued growth and success.

The journey to overcoming challenges in wind energy deployment is a testament to human ingenuity and the relentless pursuit of progress. As we continue to harness the power of the wind with ever-increasing efficiency and creativity, we move closer to a future powered by clean and sustainable energy sources. The challenges we face are not insurmountable; rather, they are opportunities for innovation and collaboration. By addressing these challenges head-on, we can unlock the full potential of wind energy and contribute to a more sustainable and resilient energy future.

The exploration of wind energy deployment is a testament to the power of innovation and the potential of renewable energy. As we continue to harness the power of the wind with ever-increasing efficiency and creativity, we move closer to a future powered by clean and sustainable energy sources. The wind, once a distant and untouchable force, is now within our grasp, offering a clean and abundant source of energy for generations to come.

Future Prospects for Wind Technology

The horizon of wind technology is brimming with possibilities, as advancements continue to reshape the landscape of renewable energy. As the world grapples with the pressing need for sustainable energy solutions, wind power stands at the forefront, offering a clean and inexhaustible resource. The future of wind technology is not just about harnessing more energy but doing so with greater efficiency, adaptability, and integration into our daily lives.

One of the most exciting prospects for wind technology is the development of next-generation turbines. These turbines are being designed to be larger, more efficient, and capable of capturing energy from a wider range of wind speeds. Innovations in materials science are playing a crucial role in this evolution. The use of lightweight composites and advanced alloys is enabling the construction of longer and more durable blades, which can capture more wind energy without compromising structural integrity. This leap in design is expected to significantly increase the power output of wind farms, making wind energy more competitive with traditional energy sources.

The concept of vertical-axis wind turbines (VAWTs) is gaining traction as a promising alternative to the conventional horizontal-axis designs. VAWTs offer several advantages, including the ability to capture wind from any direction and a smaller footprint, making them suitable for urban environments and offshore installations. Their unique design allows for quieter operation and reduced visual impact, addressing some of the common concerns associated with wind farms. As research and development in this area continue, VAWTs could become a vital component of the wind energy landscape.

Offshore wind technology is poised for significant growth, driven by the potential to harness the stronger and more consistent winds found over open waters. Floating wind turbines, in particular, are opening up new frontiers for offshore wind development. These turbines are anchored to the seabed using mooring lines, allowing them to be deployed in deeper waters where traditional fixed-bottom turbines are not feasible. The ability to access these untapped wind resources is expected to drive substantial increases in global wind energy capacity, contributing to a more diverse and resilient energy mix.

The integration of wind energy with other renewable sources is another promising avenue for future development. Hybrid systems that combine wind power with solar, hydro, or geothermal energy can provide a more consistent and reliable energy supply. By leveraging the complementary nature of different energy sources, these systems can balance fluctuations in generation and meet demand more effectively. This approach not only enhances the overall efficiency of renewable energy systems but also reduces the need for backup power from fossil fuels, contributing to a reduction in carbon emissions.

Digitalization is transforming the wind energy sector, with advanced data analytics and machine learning playing a pivotal role in optimizing turbine performance and maintenance. Real-time monitoring and predictive analytics are enabling operators to identify potential issues before they lead to costly failures, reducing downtime and extending the lifespan of wind turbines. The use of digital twins—virtual replicas of physical assets—is allowing for more accurate simulations and testing, leading to improved design and operational efficiency. As digital technologies continue to evolve, their integration into wind energy systems is expected to drive further improvements in performance and cost-effectiveness.

Energy storage solutions are becoming increasingly important for the integration of wind energy into the grid. The intermittent nature of wind power has long been a challenge, but advancements in battery technology are providing effective solutions. By storing excess energy generated during periods of high wind, these systems ensure a steady supply of electricity even when the wind is not blowing. This capability is crucial for maintaining grid stability and reliability, particularly as the share of wind energy in the overall energy mix continues to grow. The combination of wind power and energy storage is paving the way for a more flexible and resilient energy system.

The role of policy and regulation in shaping the future of wind technology cannot be overstated. Governments around the world are recognizing the importance of wind power in achieving climate goals and are implementing measures to encourage investment and development. Supportive policies, such as tax incentives, feed-in tariffs, and renewable energy certificates, are helping to level the playing field for wind energy and attract investment. International collaboration and knowledge sharing are also playing a crucial role in advancing wind technology and overcoming common challenges.

The economic benefits of wind technology are substantial, contributing to job creation, economic growth, and energy security. The construction, operation, and maintenance of wind farms require a skilled workforce, providing employment opportunities in engineering, manufacturing, and maritime industries. Additionally, the development of a robust wind energy sector can reduce reliance on imported fossil fuels, enhancing energy independence and resilience. As the global demand for clean energy continues to rise, the wind energy sector is poised to play a leading role in the transition to a sustainable energy future.

The future prospects for wind technology are a testament to the power of innovation and the potential of renewable energy. As we continue to harness the power of the wind with ever-increasing efficiency and creativity, we move closer to a future powered by clean and sustainable energy sources. The wind, once a distant and untouchable force, is now within our grasp, offering a clean and abundant source of energy for generations to come. The journey ahead is filled with challenges and opportunities, but with continued commitment and collaboration, the future of wind technology is bright and full of promise.

The Importance of Energy Storage in Renewables

Energy storage is a cornerstone in the evolution of renewable energy systems, playing a pivotal role in addressing the inherent variability and intermittency of sources like wind and solar power. As the global energy landscape shifts towards cleaner alternatives, the importance of effective energy storage solutions becomes increasingly apparent. These systems not only enhance the reliability and stability of renewable energy but also pave the way for a more resilient and sustainable energy future.

The primary challenge with renewable energy sources is their dependence on natural conditions, which can lead to fluctuations in energy production. Solar panels generate electricity only when the sun is shining, and wind turbines produce power only when the wind is blowing. This variability can lead to mismatches between energy supply and demand, posing a significant challenge for grid operators. Energy storage systems offer a solution by capturing excess energy generated during peak production periods and releasing it when demand is high or production is low. This capability ensures a consistent and reliable power supply, reducing the need for backup power from fossil fuels.

One of the most widely used energy storage technologies is the lithium-ion battery. Known for their high energy density and efficiency, lithium-ion batteries are commonly used in applications ranging from electric vehicles to grid-scale storage systems. Their ability to rapidly charge and discharge makes them ideal for balancing short-term fluctuations in renewable energy production. Advances in battery technology are continually improving their performance and reducing costs, making them an increasingly viable option for large-scale energy storage.

Pumped hydroelectric storage is another well-established energy storage solution, accounting for the majority of global energy storage capacity. This technology involves using excess electricity to pump water from a lower reservoir to a higher one. When electricity is needed, the stored water is released to flow back down through turbines, generating power. Pumped hydro storage is highly efficient and capable of storing large amounts of energy for extended periods, making it an excellent option for balancing longer-term variations in renewable energy production.

Emerging technologies, such as flow batteries and compressed air energy storage, are also showing promise in the renewable energy sector. Flow batteries use liquid electrolytes stored in external tanks, allowing for flexible and scalable energy storage solutions. Their long cycle life and ability to maintain capacity over time make them suitable for grid-scale applications. Compressed air energy storage involves using excess electricity to compress air and store it in underground caverns. When electricity is needed, the compressed air is released and expanded through turbines to generate power. These technologies offer additional options for integrating renewable energy into the grid and enhancing system resilience.

The integration of energy storage with renewable energy systems offers numerous benefits beyond grid stability. By reducing reliance on fossil fuels for backup power, energy storage contributes to a reduction in greenhouse gas emissions and air pollution. This shift towards cleaner energy sources is crucial in the fight against climate change and the pursuit of a sustainable future. Additionally, energy storage can enhance energy security by reducing dependence on imported fuels and increasing the resilience of energy systems to disruptions.

Energy storage also plays a vital role in enabling the widespread adoption of distributed energy resources, such as rooftop solar panels and small-scale wind turbines. By storing excess energy generated by these systems, homeowners and businesses can maximize their use of renewable energy and reduce their reliance on the grid. This capability not only lowers energy costs but also empowers consumers to take control of their energy usage and contribute to a more sustainable energy system.

The economic benefits of energy storage are substantial, contributing to job creation, economic growth, and energy independence. The development, manufacturing, and deployment of energy storage systems require a skilled workforce, providing employment opportunities in engineering, manufacturing, and installation. As the demand for energy storage continues to rise, the sector is poised for significant growth, driving innovation and investment in new technologies.

The role of policy and regulation in supporting the development and deployment of energy storage cannot be overstated. Governments around the world are recognizing the importance of energy storage in achieving climate goals and are implementing measures to encourage investment and development. Supportive policies, such as tax incentives, grants, and research funding, are helping to accelerate the adoption of energy storage technologies and integrate them into the energy system.

The future of energy storage is bright, with ongoing research and development driving continuous improvements in performance, cost, and scalability. As technology evolves, new and innovative storage solutions are expected to emerge, further enhancing the integration of renewable energy into the grid. The combination of renewable energy and energy storage offers a pathway to a more sustainable and resilient energy future, capable of meeting the demands of a rapidly changing world.

The journey towards a sustainable energy future is marked by challenges and opportunities, but with continued commitment and collaboration, the potential of energy storage to transform our energy systems is immense. By harnessing the power of energy storage, we can unlock the full potential of renewable energy and contribute to a cleaner, more sustainable world for generations to come.

Breakthroughs in Battery Technology

Battery technology has undergone remarkable advancements in recent years, revolutionizing the way we store and utilize energy. These breakthroughs are not only transforming the renewable energy landscape but also reshaping industries such as transportation, consumer electronics, and grid management. As the demand for efficient and sustainable energy solutions grows, the evolution of battery technology is playing a critical role in meeting these needs.

One of the most significant breakthroughs in battery technology is the development of solid-state batteries. Unlike traditional lithium-ion batteries, which use liquid electrolytes, solid-state batteries employ a solid electrolyte. This innovation offers several advantages, including higher energy density, improved safety, and longer lifespan. The absence of flammable liquid electrolytes reduces the risk of thermal runaway and fires, making solid-state batteries a safer option for a wide range of applications. Additionally, their higher energy density allows for more compact and lightweight designs, which is particularly beneficial for electric vehicles and portable electronics.

The quest for more sustainable and environmentally friendly battery materials has led to the exploration of alternatives to lithium-ion technology. Sodium-ion batteries are emerging as a promising contender, leveraging the abundance and low cost of sodium compared to lithium. While still in the developmental stage, sodium-ion batteries have the potential to offer comparable performance to lithium-ion batteries, with the added benefit of reduced environmental impact. Researchers are also investigating the use of other earth-abundant materials, such as magnesium and aluminum, to create more sustainable battery solutions.

The advancement of battery technology is closely tied to improvements in energy density, which determines how much energy a battery can store relative to its size and weight. High energy density is crucial for applications where space and weight are at a premium, such as electric vehicles and portable electronics. Recent breakthroughs in electrode materials, such as silicon anodes and lithium-metal anodes, are pushing the boundaries of energy density, enabling batteries to store more energy in a smaller footprint. These innovations are paving the way for longer-lasting electric vehicles and more powerful consumer electronics.

Fast-charging capabilities are another area where battery technology is making significant strides. The ability to rapidly charge a battery without compromising its lifespan is a key factor in the widespread adoption of electric vehicles and portable devices. Advances in electrolyte formulations and electrode designs are enabling faster charging times, reducing the inconvenience of long charging periods. This progress is particularly important for electric vehicles, where fast-charging infrastructure is essential for addressing range anxiety and promoting consumer acceptance.

The integration of smart technology and data analytics into battery management systems is enhancing the performance and reliability of modern batteries. These systems monitor and optimize battery performance in real-time, ensuring efficient energy use and prolonging battery life. Predictive analytics can identify potential issues before they lead to failures, reducing maintenance costs and downtime. The use of digital twins— virtual models of physical batteries—allows for more accurate simulations and testing, leading to improved design and operational efficiency.

Battery recycling and second-life applications are becoming increasingly important as the adoption of battery-powered technologies grows. The ability to recycle and repurpose batteries not only reduces waste but also conserves valuable materials and resources. Advances in recycling technologies are improving the efficiency and effectiveness of battery recycling processes, enabling the recovery of critical materials such as lithium, cobalt, and nickel. Second-life applications, where used batteries are repurposed for less demanding tasks, are extending the useful life of batteries and providing additional value.

The role of policy and regulation in supporting the development and deployment of advanced battery technologies is crucial. Governments around the world are recognizing the importance of batteries in achieving climate goals and are implementing measures to encourage investment and innovation. Supportive policies, such as research funding, tax incentives, and recycling mandates, are helping to accelerate the adoption of advanced battery technologies and integrate them into the energy system.

The economic impact of breakthroughs in battery technology is substantial, contributing to job creation, economic growth, and energy independence. The development, manufacturing, and deployment of advanced batteries require a skilled workforce, providing employment opportunities in engineering, manufacturing, and installation. As the demand for battery-powered technologies continues to rise, the sector is poised for significant growth, driving innovation and investment in new technologies.

The future of battery technology is bright, with ongoing research and development driving continuous improvements in performance, cost, and scalability. As technology evolves, new and innovative battery solutions are expected to emerge, further enhancing the integration of renewable energy into the grid and transforming industries such as transportation and consumer electronics. The combination of advanced battery technology and renewable energy offers a pathway to a more sustainable and resilient energy future, capable of meeting the demands of a rapidly changing world.

The journey towards a sustainable energy future is marked by challenges and opportunities, but with continued commitment and collaboration, the potential of battery technology to transform our energy systems is immense. By harnessing the power of advanced batteries, we can unlock the full potential of renewable energy and contribute to a cleaner, more sustainable world for generations to come.

Exploring Alternative Storage Solutions

The quest for efficient and sustainable energy storage solutions is driving innovation across the globe, as the need to balance energy supply and demand becomes increasingly critical. While traditional battery technologies have dominated the landscape, alternative storage solutions are emerging, offering unique advantages and addressing specific challenges associated with renewable energy integration. These innovative approaches are expanding the possibilities for energy storage, paving the way for a more resilient and adaptable energy system.

One promising alternative storage solution is the use of flywheels. Flywheel energy storage systems store energy in the form of kinetic energy by spinning a rotor at high speeds. When energy is needed, the rotor's kinetic energy is converted back into electrical energy. Flywheels offer several advantages, including high efficiency, long lifespan, and rapid response times. They are particularly well-suited for applications requiring short-duration energy storage and quick bursts of power, such as frequency regulation and grid stabilization. The ability to charge and discharge rapidly without degradation makes flywheels an attractive option for enhancing grid reliability.

Thermal energy storage is another innovative approach gaining traction in the renewable energy sector. This technology involves storing excess energy in the form of heat, which can be later converted back into electricity or used directly for heating applications. One common method of thermal energy storage is the use of molten salt, which can retain heat for extended periods. Concentrated solar power plants often utilize molten salt storage to capture and store solar energy, allowing for electricity generation even when the sun is not shining. Thermal energy storage systems offer the advantage of scalability and can be integrated with various renewable energy sources to provide a stable and reliable power supply.

Compressed air energy storage (CAES) is an alternative solution that leverages the principles of air compression and expansion to store and release energy. In a CAES system, excess electricity is used to compress air and store it in underground caverns or tanks. When electricity is needed, the compressed air is released and expanded through turbines to generate power. CAES systems are capable of storing large amounts of energy for extended periods, making them suitable for balancing long-term variations in renewable energy production. The use of existing geological formations for air storage can reduce infrastructure costs and environmental impact.

Hydrogen energy storage is an emerging technology with the potential to revolutionize the way we store and utilize energy. This approach involves using excess electricity to produce hydrogen through electrolysis, a process that splits water into hydrogen and oxygen. The hydrogen can then be stored and used as a clean fuel for electricity generation, transportation, or industrial applications. Hydrogen energy storage offers the advantage of long-term energy storage and can be integrated with renewable energy sources to create a sustainable and carbon-free energy system. The versatility of hydrogen as both an energy carrier and a fuel makes it a promising solution for addressing the challenges of renewable energy integration.

The development of advanced supercapacitors is another area of innovation in energy storage. Supercapacitors store energy through electrostatic charge rather than chemical reactions, allowing for rapid charging and discharging. They offer high power density and long cycle life, making them ideal for applications requiring quick bursts of energy, such as electric vehicles and grid stabilization. While supercapacitors have traditionally been limited by their lower energy density compared to batteries, ongoing research is focused on improving their capacity and performance. The combination of supercapacitors with other storage technologies can enhance overall system efficiency and flexibility.

The integration of alternative storage solutions with renewable energy systems offers numerous benefits beyond grid stability. By providing a diverse range of storage options, these technologies can address specific energy storage needs and optimize the use of renewable resources. This flexibility not only enhances the efficiency and reliability of energy systems but also reduces the need for backup power from fossil fuels, contributing to a reduction in greenhouse gas emissions and air pollution. The shift towards cleaner energy sources is crucial in the fight against climate change and the pursuit of a sustainable future.

The economic impact of alternative storage solutions is significant, contributing to job creation, economic growth, and energy independence. The development, manufacturing, and deployment of these technologies require a skilled workforce, providing employment opportunities in engineering, manufacturing, and installation. As the demand for energy storage continues to rise, the sector is poised for significant growth, driving innovation and investment in new technologies.

The role of policy and regulation in supporting the development and deployment of alternative storage solutions is crucial. Governments around the world are recognizing the importance of energy storage in achieving climate goals and are implementing measures to encourage investment and development. Supportive policies, such as research funding, tax incentives, and regulatory frameworks, are helping to accelerate the adoption of alternative storage technologies and integrate them into the energy system.

The future of energy storage is bright, with ongoing research and development driving continuous improvements in performance, cost, and scalability. As technology evolves, new and innovative storage solutions are expected to emerge, further enhancing the integration of renewable energy into the grid and transforming industries such as transportation and consumer electronics. The combination of alternative storage solutions and renewable energy offers a pathway to a more sustainable and resilient energy future, capable of meeting the demands of a rapidly changing world.

The journey towards a sustainable energy future is marked by challenges and opportunities, but with continued commitment and collaboration, the potential of alternative storage solutions to transform our energy systems is immense. By harnessing the power of these innovative technologies, we can unlock the full potential of renewable energy and contribute to a cleaner, more sustainable world for generations to come.

The Role of Artificial Intelligence in Energy Management

The integration of artificial intelligence (AI) into energy management systems is revolutionizing the way we produce, distribute, and consume energy. As the world transitions towards more sustainable energy sources, the complexity of managing diverse and decentralized energy systems increases. AI offers powerful tools to optimize these systems, enhance efficiency, and support the integration of renewable energy sources, ultimately contributing to a more resilient and sustainable energy future.

AI's ability to process vast amounts of data and identify patterns makes it an invaluable asset in energy management. One of the key applications of AI is in predictive analytics, where it is used to forecast energy demand and supply. By analyzing historical data and real-time information, AI algorithms can predict energy consumption patterns with remarkable accuracy. This capability allows energy providers to optimize the operation of power plants, reduce waste, and ensure a stable supply of electricity. For consumers, AI-driven demand forecasting can lead to more efficient energy use and cost savings.

The integration of renewable energy sources, such as wind and solar power, into the grid presents unique challenges due to their intermittent nature. AI plays a crucial role in addressing these challenges by enabling more effective grid management. Machine learning algorithms can analyze weather data and predict fluctuations in renewable energy production, allowing grid operators to make informed decisions about energy distribution and storage. This capability ensures that renewable energy is utilized efficiently and that the grid remains stable even when production levels vary.

AI is also transforming the way we manage energy storage systems. By optimizing the charging and discharging cycles of batteries and other storage technologies, AI can extend their lifespan and improve their performance. This optimization is particularly important for grid-scale storage systems, where efficient energy management is critical for maintaining grid stability. AI-driven energy storage management can also enhance the integration of distributed energy resources, such as rooftop solar panels, by ensuring that excess energy is stored and used effectively.

In the realm of energy efficiency, AI is making significant strides in optimizing building management systems. Smart buildings equipped with AI-driven systems can monitor and control heating, ventilation, air conditioning, lighting, and other energy-consuming processes. By analyzing data from sensors and user preferences, AI can adjust these systems in real-time to maximize energy efficiency and comfort. This capability not only reduces energy consumption and costs but also contributes to a reduction in greenhouse gas emissions.

The transportation sector is another area where AI is playing a transformative role in energy management. Electric vehicles (EVs) are becoming increasingly popular, and AI is helping to optimize their charging and energy use. Smart charging systems powered by AI can determine the best times to charge EVs based on energy prices, grid demand, and renewable energy availability. This optimization ensures that EVs are charged efficiently and cost-effectively, while also supporting the integration of renewable energy into the grid.

AI is also being used to enhance the efficiency and reliability of energy infrastructure. Predictive maintenance, powered by AI, allows energy providers to identify potential equipment failures before they occur. By analyzing data from sensors and historical maintenance records, AI algorithms can predict when equipment is likely to fail and schedule maintenance accordingly. This proactive approach reduces downtime, extends the lifespan of equipment, and lowers maintenance costs.

The role of AI in energy management extends beyond optimization and efficiency. AI is also being used to enhance cybersecurity in energy systems. As energy infrastructure becomes more digital and interconnected, the risk of cyberattacks increases. AI-driven cybersecurity solutions can detect and respond to threats in real-time, protecting critical energy infrastructure from potential attacks. This capability is essential for ensuring the security and resilience of energy systems in an increasingly digital world.

The economic impact of AI in energy management is substantial, contributing to job creation, economic growth, and energy independence. The development and deployment of AI-driven energy solutions require a skilled workforce, providing employment opportunities in data science, engineering, and technology. As the demand for AI-driven energy management continues to rise, the sector is poised for significant growth, driving innovation and investment in new technologies.

The role of policy and regulation in supporting the integration of AI into energy management is crucial. Governments around the world are recognizing the importance of AI in achieving climate goals and are implementing measures to encourage investment and development. Supportive policies, such as research funding, tax incentives, and regulatory frameworks, are helping to accelerate the adoption of AI-driven energy solutions and integrate them into the energy system.

The future of AI in energy management is bright, with ongoing research and development driving continuous improvements in performance, cost, and scalability. As technology evolves, new and innovative AI-driven solutions are expected to emerge, further enhancing the integration of renewable energy into the grid and transforming industries such as transportation and building management. The combination of AI and renewable energy offers a pathway to a more sustainable and resilient energy future, capable of meeting the demands of a rapidly changing world.

The journey towards a sustainable energy future is marked by challenges and opportunities, but with continued commitment and collaboration, the potential of AI to transform our energy systems is immense. By harnessing the power of AI, we can unlock the full potential of renewable energy and contribute to a cleaner, more sustainable world for generations to come.

Real-World Applications of Advanced Storage Systems

Advanced storage systems are becoming increasingly integral to the modern energy landscape, offering solutions that address the challenges of integrating renewable energy sources into the grid. These systems are not just theoretical constructs; they are being implemented in real-world scenarios across various sectors, demonstrating their potential to revolutionize energy management and consumption. By examining these applications, we can gain a deeper understanding of how advanced storage systems are shaping the future of energy.

One of the most prominent applications of advanced storage systems is in grid stabilization and management. As renewable energy sources like wind and solar become more prevalent, the grid faces challenges related to their intermittent nature. Energy storage systems, such as large-scale batteries and pumped hydro storage, are being deployed to balance supply and demand, ensuring a stable and reliable power supply. These systems store excess energy generated during periods of high production and release it when demand peaks or production dips. This capability not only enhances grid stability but also reduces the need for fossil fuel-based backup power, contributing to a reduction in carbon emissions.

In the transportation sector, advanced storage systems are playing a crucial role in the electrification of vehicles. Electric vehicles (EVs) rely on high-performance batteries to provide the energy needed for propulsion. The development of lithium-ion and solid-state batteries has significantly improved the range, efficiency, and charging times of EVs, making them a viable alternative to traditional internal combustion engine vehicles. As battery technology continues to advance, we can expect further improvements in EV performance, driving wider adoption and reducing the transportation sector's carbon footprint.

Residential and commercial buildings are also benefiting from the integration of advanced storage systems. Homeowners and businesses are increasingly installing battery storage systems to complement their solar panel installations. These systems allow users to store excess solar energy generated during the day and use it during the evening or on cloudy days, maximizing the use of renewable energy and reducing reliance on the grid. In addition to cost savings, this approach enhances energy independence and resilience, particularly in areas prone to power outages.

Microgrids are another real-world application of advanced storage systems, providing localized energy solutions that can operate independently or in conjunction with the main grid. Microgrids are particularly valuable in remote or underserved areas where access to reliable electricity is limited. By integrating renewable energy sources with energy storage, microgrids can provide a stable and sustainable power supply, supporting economic development and improving quality of life. In urban areas, microgrids offer a way to enhance grid resilience and reduce the impact of power outages, making them an attractive option for critical infrastructure such as hospitals and data centers.

The industrial sector is leveraging advanced storage systems to optimize energy use and reduce costs. Energy-intensive industries, such as manufacturing and mining, are adopting storage solutions to manage peak demand and improve energy efficiency. By storing energy during periods of low demand and using it during peak times, these industries can reduce their energy costs and minimize their environmental impact. Additionally, energy storage systems can provide backup power during outages, ensuring continuity of operations and reducing downtime.

Renewable energy projects are increasingly incorporating advanced storage systems to enhance their viability and competitiveness. Wind and solar farms are integrating battery storage to smooth out fluctuations in energy production and provide a more consistent power supply. This integration allows renewable energy projects to participate in ancillary services markets, providing grid support functions such as frequency regulation and voltage control. By enhancing the reliability and predictability of renewable energy, storage systems are helping to accelerate the transition to a cleaner energy future.

The role of advanced storage systems in disaster recovery and emergency response is gaining recognition. In the aftermath of natural disasters, energy storage systems can provide critical backup power to support emergency services and aid in recovery efforts. Portable storage solutions, such as battery packs and mobile microgrids, can be deployed quickly to provide electricity in areas where the grid has been damaged. This capability is essential for maintaining communication, powering medical equipment, and supporting relief operations, highlighting the importance of energy storage in enhancing resilience and preparedness.

The economic impact of advanced storage systems is significant, driving job creation and economic growth. The development, manufacturing, and deployment of these systems require a skilled workforce, providing employment opportunities in engineering, manufacturing, and installation. As the demand for energy storage continues to rise, the sector is poised for significant growth, attracting investment and fostering innovation in new technologies.

Policy and regulation play a crucial role in supporting the deployment of advanced storage systems. Governments around the world are recognizing the importance of energy storage in achieving climate goals and are implementing measures to encourage investment and development. Supportive policies, such as research funding, tax incentives, and regulatory frameworks, are helping to accelerate the adoption of advanced storage technologies and integrate them into the energy system.

The future of advanced storage systems is bright, with ongoing research and development driving continuous improvements in performance, cost, and scalability. As technology evolves, new and innovative storage solutions are expected to emerge, further enhancing the integration of renewable energy into the grid and transforming industries such as transportation and building management. The combination of advanced storage systems and renewable energy offers a pathway to a more sustainable and resilient energy future, capable of meeting the demands of a rapidly changing world.

The journey towards a sustainable energy future is marked by challenges and opportunities, but with continued commitment and collaboration, the potential of advanced storage systems to transform our energy systems is immense. By harnessing the power of these innovative technologies, we can unlock the full potential of renewable energy and contribute to a cleaner, more sustainable world for generations to come.

Chapter 4: Hydrogen and Fuel Cell Innovations

Understanding Hydrogen as a Clean Energy Source

Hydrogen, often hailed as the fuel of the future, is gaining momentum as a clean energy source with the potential to revolutionize the way we produce and consume energy. Its versatility and abundance make it an attractive option for addressing the challenges of climate change and energy security. Understanding the role of hydrogen in the energy landscape requires a closer look at its production methods, applications, and the benefits it offers as a sustainable energy solution.

Hydrogen is the most abundant element in the universe, yet it rarely exists in its pure form on Earth. Instead, it is typically found in compounds such as water and hydrocarbons. To harness hydrogen as an energy source, it must first be extracted from these compounds through various production methods. The most common method is steam methane reforming (SMR), which involves reacting natural gas with steam to produce hydrogen and carbon dioxide. While SMR is currently the most cost-effective method, it is not entirely clean due to the carbon emissions associated with the process.

To address the environmental impact of hydrogen production, researchers are exploring alternative methods that produce hydrogen with minimal or no carbon emissions. One promising approach is electrolysis, which uses electricity to split water into hydrogen and oxygen. When powered by renewable energy sources such as wind or solar, electrolysis can produce "green hydrogen," a truly clean and sustainable energy carrier.

Although the cost of electrolysis is currently higher than SMR, advancements in technology and economies of scale are expected to make green hydrogen more competitive in the future.

Hydrogen's versatility as an energy carrier is one of its most significant advantages. It can be used in various applications, ranging from transportation and power generation to industrial processes and heating. In the transportation sector, hydrogen fuel cells offer a clean alternative to traditional internal combustion engines. Fuel cell vehicles (FCVs) convert hydrogen into electricity, emitting only water vapor as a byproduct. This technology is particularly well-suited for heavy-duty vehicles such as buses and trucks, where the energy density and quick refueling capabilities of hydrogen provide a practical solution for long-range travel.

In power generation, hydrogen can be used as a fuel for gas turbines, providing a flexible and low-emission option for electricity production. Hydrogen-fired power plants can complement renewable energy sources by providing backup power during periods of low wind or solar output. Additionally, hydrogen can be stored and transported over long distances, making it an ideal solution for balancing energy supply and demand across regions.

The industrial sector is another area where hydrogen can play a transformative role. Industries such as steel, cement, and chemicals are among the largest emitters of carbon dioxide, and hydrogen offers a pathway to decarbonize these processes. For example, hydrogen can be used as a reducing agent in steel production, replacing carbon-intensive coke and significantly reducing emissions. Similarly, hydrogen can be used as a feedstock for producing ammonia and methanol, essential chemicals for fertilizers and plastics, respectively.

Hydrogen's potential as a clean energy source extends beyond its direct applications. It can also serve as a means of energy storage, addressing one of the key challenges of integrating renewable energy into the grid. By converting excess electricity from renewable sources into hydrogen through electrolysis, energy can be stored for later use or transported to areas with high demand. This capability enhances grid stability and resilience, supporting the transition to a more sustainable energy system.

Despite its potential, the widespread adoption of hydrogen as a clean energy source faces several challenges. The cost of production, infrastructure development, and regulatory frameworks are among the key barriers that need to be addressed. The production of green hydrogen, while environmentally friendly, is currently more expensive than conventional methods. However, ongoing research and development efforts are focused on reducing costs and improving the efficiency of electrolysis and other clean production technologies.

Infrastructure development is another critical factor in the hydrogen economy. The transportation and storage of hydrogen require specialized infrastructure, including pipelines, refueling stations, and storage facilities. Building this infrastructure will require significant investment and collaboration between governments, industry, and other stakeholders. Policymakers play a crucial role in supporting the development of hydrogen infrastructure through incentives, funding, and regulatory frameworks that encourage investment and innovation.

The role of policy and regulation in promoting hydrogen as a clean energy source cannot be overstated. Governments around the world are recognizing the importance of hydrogen in achieving climate goals and are implementing measures to support its development and deployment. National hydrogen strategies, research funding, and public-private partnerships are helping to accelerate the growth of the hydrogen economy and integrate it into the broader energy system.

The economic impact of hydrogen as a clean energy source is substantial, offering opportunities for job creation, economic growth, and energy independence. The development and deployment of hydrogen technologies require a skilled workforce, providing employment opportunities in engineering, manufacturing, and installation. As the demand for hydrogen continues to rise, the sector is poised for significant growth, driving innovation and investment in new technologies.

The future of hydrogen as a clean energy source is promising, with ongoing research and development driving continuous improvements in performance, cost, and scalability. As technology evolves, new and innovative hydrogen solutions are expected to emerge, further enhancing the integration of renewable energy into the grid and transforming industries such as transportation and manufacturing. The combination of hydrogen and renewable energy offers a pathway to a more sustainable and resilient energy future, capable of meeting the demands of a rapidly changing world.

The journey towards a sustainable energy future is marked by challenges and opportunities, but with continued commitment and collaboration, the potential of hydrogen to transform our energy systems is immense. By harnessing the power of hydrogen, we can unlock the full potential of renewable energy and contribute to a cleaner, more sustainable world for generations to come.

Advances in Fuel Cell Technology

Fuel cell technology has emerged as a promising solution in the quest for cleaner and more efficient energy systems. As the world grapples with the challenges of climate change and the need for sustainable energy sources, fuel cells offer a versatile and environmentally friendly alternative to traditional combustion-based power generation. Recent advances in fuel cell technology have significantly enhanced their performance, efficiency, and applicability across various sectors, paving the way for a broader adoption of this innovative energy solution.

At the heart of fuel cell technology is the electrochemical process that converts chemical energy into electrical energy. Unlike conventional combustion engines, fuel cells generate electricity through a chemical reaction between hydrogen and oxygen, producing only water and heat as byproducts. This clean and efficient process makes fuel cells an attractive option for reducing greenhouse gas emissions and air pollution. The versatility of fuel cells allows them to be used in a wide range of applications, from powering vehicles and portable devices to providing backup power for buildings and industrial facilities.

One of the most significant advances in fuel cell technology is the development of proton exchange membrane fuel cells (PEMFCs). These fuel cells operate at relatively low temperatures and offer high power density, making them ideal for applications such as transportation and portable electronics.

Recent improvements in PEMFCs have focused on enhancing their durability and reducing costs. Researchers have made strides in developing more robust and efficient catalysts, which are essential for facilitating the chemical reactions within the fuel cell. By reducing the reliance on expensive materials like platinum, these advancements have the potential to lower the overall cost of PEMFCs and make them more accessible for widespread use.

Solid oxide fuel cells (SOFCs) represent another area of significant advancement in fuel cell technology. Operating at high temperatures, SOFCs are well-suited for stationary power generation and industrial applications. Their high efficiency and fuel flexibility allow them to utilize a variety of fuels, including natural gas and biogas, making them a versatile option for different energy needs. Recent developments in SOFCs have focused on improving their thermal stability and reducing startup times, addressing some of the challenges associated with their high operating temperatures. These improvements are enhancing the viability of SOFCs for both large-scale power generation and distributed energy systems.

The transportation sector is one of the most promising areas for the application of fuel cell technology. Fuel cell vehicles (FCVs) offer a clean and efficient alternative to traditional internal combustion engine vehicles, with the added benefit of quick refueling times compared to battery electric vehicles. Advances in fuel cell technology have led to significant improvements in the range and performance of FCVs, making them a more competitive option in the automotive market. The development of hydrogen refueling infrastructure is also progressing, with governments and private companies investing in the expansion of refueling stations to support the growing number of FCVs on the road.

In addition to passenger vehicles, fuel cells are being used to power buses, trucks, and even trains, offering a sustainable solution for reducing emissions in the transportation sector. The high energy density and rapid refueling capabilities of fuel cells make them particularly well-suited for heavy-duty and long-range applications, where battery electric solutions may be less practical. As fuel cell technology continues to advance, we can expect to see an increasing number of commercial and public transportation systems adopting this clean energy solution.

Fuel cells are also making inroads into the realm of portable and backup power solutions. Their ability to provide reliable and efficient power makes them an attractive option for remote and off-grid applications, as well as for emergency backup power in critical infrastructure. Advances in fuel cell technology have led to the development of compact and lightweight systems that can be easily deployed in a variety of settings. These portable fuel cells are being used to power everything from remote telecommunications equipment to military operations, providing a dependable source of energy in challenging environments.

The integration of fuel cells into renewable energy systems is another area of growing interest. By coupling fuel cells with renewable energy sources such as solar and wind, it is possible to create hybrid systems that offer both clean energy generation and reliable power supply. Fuel cells can provide backup power during periods of low renewable energy production, ensuring a stable and continuous energy supply. This integration enhances the overall efficiency and resilience of renewable energy systems, supporting the transition to a more sustainable energy future.

Despite the significant advances in fuel cell technology, several challenges remain in achieving widespread adoption. The cost of fuel cell systems, particularly for transportation applications, remains a barrier to entry for many consumers. However, ongoing research and development efforts are focused on reducing costs through improved materials and manufacturing processes. Additionally, the development of hydrogen production and distribution infrastructure is critical for supporting the growth of fuel cell technology. Policymakers and industry stakeholders play a crucial role in facilitating this development through supportive policies, incentives, and investment in infrastructure projects.

The economic impact of fuel cell technology is substantial, offering opportunities for job creation and economic growth. The development, manufacturing, and deployment of fuel cell systems require a skilled workforce, providing employment opportunities in engineering, manufacturing, and installation. As the demand for fuel cell technology continues to rise, the sector is poised for significant growth, driving innovation and investment in new technologies.

The future of fuel cell technology is promising, with ongoing research and development driving continuous improvements in performance, cost, and scalability. As technology evolves, new and innovative fuel cell solutions are expected to emerge, further enhancing the integration of clean energy into the grid and transforming industries such as transportation and power generation. The combination of fuel cells and renewable energy offers a pathway to a more sustainable and resilient energy future, capable of meeting the demands of a rapidly changing world.

The journey towards a sustainable energy future is marked by challenges and opportunities, but with continued commitment and collaboration, the potential of fuel cell technology to transform our energy systems is immense. By harnessing the power of fuel cells, we can unlock the full potential of clean energy and contribute to a cleaner, more sustainable world for generations to come.

The Potential of Hydrogen in Transportation

The transportation sector, a significant contributor to global greenhouse gas emissions, is undergoing a transformative shift towards cleaner and more sustainable energy sources. Among the promising alternatives, hydrogen stands out as a versatile and efficient energy carrier with the potential to revolutionize transportation. Its unique properties offer solutions to some of the most pressing challenges faced by the industry, including the need for zero-emission vehicles, energy security, and the reduction of reliance on fossil fuels.

Hydrogen's appeal in transportation lies in its ability to power fuel cells, which convert chemical energy directly into electricity through an electrochemical reaction with oxygen. This process emits only water vapor and heat, making it an environmentally friendly alternative to internal combustion engines. Fuel cell vehicles (FCVs) are at the forefront of this hydrogen revolution, offering a clean and efficient mode of transportation that rivals traditional vehicles in performance and convenience.

One of the key advantages of hydrogen-powered FCVs is their quick refueling time, which is comparable to that of gasoline vehicles. This feature addresses one of the major limitations of battery electric vehicles (BEVs), which often require extended charging periods. For consumers accustomed to the convenience of rapid refueling, FCVs present an attractive option that does not compromise on efficiency or environmental benefits. As hydrogen refueling infrastructure continues to expand, the practicality of FCVs is expected to increase, making them a viable choice for a broader range of consumers.

The range of FCVs is another compelling factor in their favor. Hydrogen fuel cells offer a higher energy density compared to batteries, allowing vehicles to travel longer distances on a single tank of hydrogen. This characteristic is particularly advantageous for long-haul transportation and commercial fleets, where extended range and minimal downtime are critical. As a result, hydrogen is gaining traction in sectors such as trucking, buses, and even trains, where the demand for sustainable and efficient transportation solutions is high.

In the realm of public transportation, hydrogen-powered buses are emerging as a sustainable alternative to diesel-powered counterparts. Cities around the world are adopting hydrogen buses to reduce urban air pollution and greenhouse gas emissions. These buses offer the same operational flexibility as traditional buses, with the added benefit of zero tailpipe emissions. The deployment of hydrogen buses not only contributes to cleaner air but also serves as a visible commitment to sustainable urban mobility.

The potential of hydrogen in transportation extends beyond road vehicles. The maritime and aviation industries are also exploring hydrogen as a means to decarbonize their operations. In shipping, hydrogen fuel cells can power vessels with minimal environmental impact, offering a solution to the industry's significant carbon footprint. Similarly, the aviation sector is investigating the use of hydrogen as a fuel for aircraft, with the goal of achieving zero-emission flights. While these applications are still in the early stages of development, they represent exciting opportunities for hydrogen to play a transformative role in global transportation.

The integration of hydrogen into transportation is not without its challenges. The production, storage, and distribution of hydrogen require significant investment and infrastructure development. Currently, most hydrogen is produced through steam methane reforming, a process that emits carbon dioxide. To fully realize hydrogen's potential as a clean energy source, the industry must transition to low-carbon production methods, such as electrolysis powered by renewable energy. This shift will require collaboration between governments, industry, and researchers to develop cost-effective and scalable solutions.

Storage and distribution are also critical components of the hydrogen supply chain. Hydrogen's low energy density in its gaseous form necessitates compression or liquefaction for efficient storage and transport. Developing the necessary infrastructure, including pipelines, refueling stations, and storage facilities, is essential for supporting the widespread adoption of hydrogen in transportation. Policymakers play a crucial role in facilitating this development through supportive policies, incentives, and investment in infrastructure projects.

The economic impact of hydrogen in transportation is substantial, offering opportunities for job creation and economic growth. The development, manufacturing, and deployment of hydrogen technologies require a skilled workforce, providing employment opportunities in engineering, manufacturing, and installation. As the demand for hydrogen-powered transportation continues to rise, the sector is poised for significant growth, driving innovation and investment in new technologies.

The role of policy and regulation in promoting hydrogen as a transportation fuel cannot be overstated. Governments around the world are recognizing the importance of hydrogen in achieving climate goals and are implementing measures to support its development and deployment. National hydrogen strategies, research funding, and public-private partnerships are helping to accelerate the growth of the hydrogen economy and integrate it into the broader transportation system.

The future of hydrogen in transportation is promising, with ongoing research and development driving continuous improvements in performance, cost, and scalability. As technology evolves, new and innovative hydrogen solutions are expected to emerge, further enhancing the integration of clean energy into the transportation sector and transforming industries such as shipping and aviation. The combination of hydrogen and renewable energy offers a pathway to a more sustainable and resilient transportation future, capable of meeting the demands of a rapidly changing world.

The journey towards a sustainable transportation future is marked by challenges and opportunities, but with continued commitment and collaboration, the potential of hydrogen to transform our transportation systems is immense. By harnessing the power of hydrogen, we can unlock the full potential of clean energy and contribute to a cleaner, more sustainable world for generations to come.

Challenges and Opportunities in Hydrogen Adoption

Hydrogen, with its potential to serve as a clean and versatile energy carrier, stands at the forefront of the global transition towards sustainable energy systems. However, the path to widespread hydrogen adoption is fraught with challenges that must be navigated to unlock its full potential. At the same time, these challenges present unique opportunities for innovation, collaboration, and growth within the energy sector. Understanding these dynamics is crucial for stakeholders aiming to harness hydrogen's capabilities in addressing climate change and energy security.

One of the primary challenges in hydrogen adoption is the current cost of production. Most hydrogen today is produced through steam methane reforming (SMR), a process that, while cost-effective, results in significant carbon emissions. To mitigate these emissions, the industry is exploring carbon capture and storage (CCS) technologies, which can reduce the carbon footprint of SMR. However, the implementation of CCS adds to the overall cost, making hydrogen less competitive compared to fossil fuels. The development of green hydrogen, produced via electrolysis using renewable energy, offers a cleaner alternative but remains more expensive due to the high costs of renewable electricity and electrolyzer technology. Reducing these costs through technological advancements and economies of scale is essential for making hydrogen a viable option for widespread use.

Infrastructure development is another critical hurdle in the hydrogen adoption journey. The transportation, storage, and distribution of hydrogen require specialized infrastructure, including pipelines, refueling stations, and storage facilities. Building this infrastructure demands substantial investment and coordination among governments, industry, and other stakeholders. The lack of a comprehensive hydrogen infrastructure network limits the accessibility and convenience of hydrogen as an energy source, particularly in the transportation sector. Addressing this challenge requires strategic planning and investment to create a robust and interconnected hydrogen infrastructure that can support the growing demand.

Safety concerns also play a significant role in the adoption of hydrogen. As a highly flammable gas, hydrogen poses unique safety challenges that must be addressed to ensure its safe handling and use. Advances in materials science and engineering have led to the development of safer storage and transportation solutions, such as high-strength composite tanks and leak detection systems. Public perception and acceptance of hydrogen as a safe energy source are crucial for its widespread adoption. Education and awareness campaigns can help dispel myths and build confidence in hydrogen technologies, paving the way for broader acceptance.

Despite these challenges, the opportunities presented by hydrogen adoption are substantial. Hydrogen's versatility as an energy carrier allows it to be used across various sectors, from transportation and power generation to industrial processes and heating. This versatility opens up new markets and applications, driving innovation and economic growth. The development of hydrogen technologies can create jobs and stimulate investment in research and development, fostering a vibrant and dynamic energy sector.

In the transportation sector, hydrogen offers a pathway to decarbonize road, rail, maritime, and aviation transport. Fuel cell vehicles (FCVs) provide a clean and efficient alternative to internal combustion engines, with the added benefit of quick refueling times and long ranges. The deployment of hydrogen-powered buses, trucks, and trains is gaining momentum, supported by government incentives and investments in refueling infrastructure. The maritime and aviation industries are also exploring hydrogen as a means to reduce their carbon footprints, with promising developments in hydrogen-powered ships and aircraft.

The integration of hydrogen into power generation presents another significant opportunity. Hydrogen can be used as a fuel for gas turbines, providing a flexible and low-emission option for electricity production. It can also serve as a means of energy storage, addressing the intermittency challenges of renewable energy sources like wind and solar. By converting excess renewable electricity into hydrogen through electrolysis, energy can be stored and used when needed, enhancing grid stability and resilience.

The industrial sector stands to benefit from hydrogen adoption as well. Industries such as steel, cement, and chemicals are among the largest emitters of carbon dioxide, and hydrogen offers a pathway to decarbonize these processes. For example, hydrogen can replace carbon-intensive coke in steel production, significantly reducing emissions. Similarly, hydrogen can be used as a feedstock for producing ammonia and methanol, essential chemicals for fertilizers and plastics, respectively. The transition to hydrogen-based industrial processes can drive innovation and competitiveness, positioning industries for a low-carbon future.

Policy and regulation play a crucial role in supporting hydrogen adoption. Governments around the world are recognizing the importance of hydrogen in achieving climate goals and are implementing measures to encourage its development and deployment. National hydrogen strategies, research funding, and public-private partnerships are helping to accelerate the growth of the hydrogen economy and integrate it into the broader energy system. Supportive policies, such as tax incentives and regulatory frameworks, can create a favorable environment for investment and innovation in hydrogen technologies.

International collaboration is also essential for advancing hydrogen adoption. The global nature of energy markets and supply chains necessitates cooperation among countries to develop standards, share best practices, and coordinate infrastructure development. Collaborative efforts can drive down costs, enhance technology transfer, and create a more integrated and resilient global hydrogen economy.

The future of hydrogen adoption is promising, with ongoing research and development driving continuous improvements in performance, cost, and scalability. As technology evolves, new and innovative hydrogen solutions are expected to emerge, further enhancing the integration of clean energy into the grid and transforming industries such as transportation and manufacturing. The combination of hydrogen and renewable energy offers a pathway to a more sustainable and resilient energy future, capable of meeting the demands of a rapidly changing world.

The journey towards a sustainable energy future is marked by challenges and opportunities, but with continued commitment and collaboration, the potential of hydrogen to transform our energy systems is immense. By harnessing the power of hydrogen, we can unlock the full potential of clean energy and contribute to a cleaner, more sustainable world for generations to come.

Pioneering Projects in Hydrogen Energy

Pioneering projects in hydrogen energy are setting the stage for a transformative shift in how we produce, store, and utilize energy. These initiatives, scattered across the globe, are not only pushing the boundaries of technology but also demonstrating the practical viability of hydrogen as a cornerstone of a sustainable energy future. By examining these trailblazing projects, we can gain insights into the innovative approaches being employed to harness hydrogen's potential and the impact they are having on the energy landscape.

One of the most ambitious hydrogen projects is the European Union's Hydrogen Strategy, which aims to establish a comprehensive hydrogen economy across the continent. This initiative seeks to integrate hydrogen into various sectors, including transportation, industry, and power generation, with the goal of achieving climate neutrality by 2050. Central to this strategy is the development of green hydrogen, produced through electrolysis powered by renewable energy sources. The EU is investing heavily in research and infrastructure to scale up hydrogen production and create a robust supply chain. By fostering collaboration among member states and industry stakeholders, the EU is positioning itself as a global leader in hydrogen innovation.

In Japan, the Fukushima Hydrogen Energy Research Field (FH2R) stands as a testament to the country's commitment to hydrogen as a clean energy solution. Located in Namie Town, Fukushima Prefecture, this facility is one of the world's largest hydrogen production sites, capable of producing up to 1,200 cubic meters of hydrogen per hour. The project utilizes renewable energy from a nearby solar power plant to power its electrolysis process, demonstrating the potential for integrating hydrogen production with renewable energy sources. FH2R serves as a model for sustainable energy systems and plays a crucial role in Japan's broader hydrogen strategy, which aims to establish a hydrogen-based society by 2050.

Australia is also making significant strides in hydrogen energy, with the Hydrogen Energy Supply Chain (HESC) project leading the charge. This ambitious initiative aims to produce hydrogen from brown coal in the Latrobe Valley, Victoria, and transport it to Japan in liquefied form. The project incorporates carbon capture and storage (CCS) technology to mitigate emissions, making it a low-carbon solution for hydrogen production. HESC represents a collaboration between Australian and Japanese companies and governments, highlighting the importance of international partnerships in advancing hydrogen technology. By demonstrating the feasibility of hydrogen export, this project is paving the way for Australia to become a major player in the global hydrogen market.

In the United States, the Los Angeles Department of Water and Power (LADWP) is spearheading the Intermountain Power Project (IPP) Renewal, a groundbreaking initiative to transition a coal-fired power plant to a hydrogen-powered facility. Located in Utah, the IPP Renewal project aims to replace coal with a blend of natural gas and hydrogen, eventually transitioning to 100% hydrogen by 2045. This project is significant not only for its scale but also for its potential to serve as a model for other power plants seeking to decarbonize. By leveraging existing infrastructure and integrating hydrogen into the power grid, the IPP Renewal project is demonstrating the viability of hydrogen as a clean energy source for large-scale power generation.

The Hydrogen Valleys initiative, supported by the Fuel Cells and Hydrogen Joint Undertaking (FCH JU), is another pioneering effort to create integrated hydrogen ecosystems across Europe. These "valleys" are regions where hydrogen production, distribution, and consumption are concentrated, creating a self-sustaining hydrogen economy. By connecting various sectors, such as transportation, industry, and heating, Hydrogen Valleys aim to maximize the efficiency and impact of hydrogen technologies. This initiative is fostering collaboration among local governments, industry, and research institutions, driving innovation and investment in hydrogen infrastructure and applications.

In the Middle East, the NEOM project in Saudi Arabia is setting a new benchmark for hydrogen energy. As part of the country's Vision 2030 plan, NEOM is a futuristic city that aims to be powered entirely by renewable energy, with hydrogen playing a central role. The project includes the construction of a large-scale green hydrogen plant, which will produce hydrogen through electrolysis using wind and solar power. NEOM's ambitious vision for a sustainable city powered by hydrogen is attracting global attention and investment, positioning Saudi Arabia as a leader in the transition to clean energy.

These pioneering projects are not only advancing hydrogen technology but also addressing some of the key challenges associated with its adoption. By demonstrating the feasibility and scalability of hydrogen solutions, these initiatives are helping to build confidence in hydrogen as a viable energy source. They are also driving down costs through economies of scale and technological innovation, making hydrogen more competitive with traditional energy sources.

The success of these projects hinges on collaboration among governments, industry, and research institutions. Public-private partnerships are playing a crucial role in advancing hydrogen technology, providing the necessary funding, expertise, and infrastructure to bring these projects to fruition. Policymakers are also instrumental in creating a supportive regulatory environment that encourages investment and innovation in hydrogen energy.

As these pioneering projects continue to evolve, they are paving the way for a broader adoption of hydrogen technologies. By showcasing the potential of hydrogen to transform energy systems, these initiatives are inspiring new projects and investments around the world. The lessons learned from these trailblazing efforts will be invaluable in guiding the development of future hydrogen projects and shaping the global energy landscape.

The journey towards a hydrogen-powered future is marked by challenges and opportunities, but with continued commitment and collaboration, the potential of hydrogen to transform our energy systems is immense. By harnessing the power of hydrogen, we can unlock the full potential of clean energy and contribute to a cleaner, more sustainable world for generations to come.

The Evolution of Smart Grid Technology

Smart grid technology represents a significant evolution in the way electricity is generated, distributed, and consumed. This modernized electrical grid leverages digital communication technology to improve the efficiency, reliability, and sustainability of electricity services. As the world increasingly turns to renewable energy sources and seeks to reduce carbon emissions, the development and implementation of smart grids have become crucial in meeting these goals. The journey of smart grid technology is marked by innovation, adaptation, and the integration of cutting-edge technologies that are reshaping the energy landscape.

The traditional electrical grid, while effective in its time, was designed for a one-way flow of electricity from centralized power plants to consumers. This system, however, is ill-suited to accommodate the complexities of modern energy demands, which include the integration of distributed energy resources (DERs) such as solar panels and wind turbines. Smart grids address these challenges by enabling a two-way flow of electricity and information, allowing for real-time monitoring and management of the grid. This capability is essential for balancing supply and demand, integrating renewable energy sources, and enhancing grid resilience.

One of the key components of smart grid technology is advanced metering infrastructure (AMI), which includes smart meters that provide real-time data on electricity consumption. These meters enable utilities to monitor energy usage more accurately and offer consumers detailed insights into their energy consumption patterns. With this information, consumers can make informed decisions about their energy use, potentially reducing their electricity bills and contributing to energy conservation efforts. Additionally, smart meters facilitate demand response programs, where consumers are incentivized to reduce their energy usage during peak demand periods, helping to alleviate stress on the grid.

The integration of renewable energy sources into the grid is a central focus of smart grid technology. Unlike traditional power plants, renewable energy sources such as solar and wind are intermittent and decentralized, posing challenges for grid stability and reliability. Smart grids employ advanced forecasting and energy management systems to predict and manage the variability of renewable energy generation. By using real-time data and predictive analytics, grid operators can optimize the use of renewable energy, ensuring a stable and reliable power supply while minimizing reliance on fossil fuels.

Energy storage systems, such as batteries, are another critical component of smart grids. These systems store excess energy generated during periods of low demand and release it during peak demand, smoothing out fluctuations in energy supply and demand. The integration of energy storage into smart grids enhances grid flexibility and resilience, enabling the efficient use of renewable energy and reducing the need for backup power from fossil fuel-based sources. As battery technology continues to advance, the cost and efficiency of energy storage systems are expected to improve, further supporting the growth of smart grids.

The evolution of smart grid technology is also characterized by the increasing use of automation and control systems. These systems enable grid operators to monitor and manage the grid in real-time, quickly identifying and responding to issues such as outages or equipment failures. Automated systems can reroute electricity around problem areas, minimizing disruptions and improving grid reliability. This capability is particularly important in the face of extreme weather events, which are becoming more frequent and severe due to climate change. By enhancing grid resilience, smart grids can help mitigate the impact of such events on electricity services.

Cybersecurity is a critical consideration in the development of smart grid technology. As the grid becomes more interconnected and reliant on digital communication, it becomes vulnerable to cyber threats. Protecting the grid from cyberattacks is essential to ensure the security and reliability of electricity services. Smart grid technology incorporates advanced cybersecurity measures, including encryption, authentication, and intrusion detection systems, to safeguard the grid against potential threats. Ongoing research and collaboration among industry stakeholders are essential to staying ahead of evolving cyber threats and ensuring the continued security of smart grids.

The deployment of smart grid technology offers numerous benefits for utilities, consumers, and the environment. For utilities, smart grids provide greater operational efficiency, reducing costs associated with energy generation, transmission, and distribution. By optimizing the use of existing infrastructure and minimizing energy losses, utilities can improve their bottom line while delivering reliable and affordable electricity services. For consumers, smart grids offer greater control over energy usage and the potential for cost savings through demand response programs and energy efficiency measures.

From an environmental perspective, smart grids play a crucial role in reducing carbon emissions and promoting the use of renewable energy. By enabling the integration of distributed energy resources and optimizing energy use, smart grids contribute to a cleaner and more sustainable energy system. The reduction in reliance on fossil fuels not only decreases greenhouse gas emissions but also improves air quality, benefiting public health and the environment.

The evolution of smart grid technology is an ongoing process, driven by advances in digital communication, data analytics, and energy management systems. As technology continues to evolve, new and innovative solutions are expected to emerge, further enhancing the capabilities and benefits of smart grids. The transition to a smart grid is a complex and multifaceted endeavor, requiring collaboration among utilities, regulators, technology providers, and consumers. By working together, stakeholders can overcome the challenges associated with smart grid deployment and unlock the full potential of this transformative technology.

The future of smart grid technology is promising, with the potential to revolutionize the way we produce, distribute, and consume electricity. As the world moves towards a more sustainable energy future, smart grids will play a central role in enabling the integration of renewable energy, enhancing grid resilience, and empowering consumers to take control of their energy use. By embracing smart grid technology, we can create a more efficient, reliable, and sustainable energy system that meets the needs of future generations.

Enhancing Grid Resilience and Reliability

Grid resilience and reliability have become paramount in the face of increasing energy demands, climate change, and the integration of renewable energy sources. As the backbone of modern society, the electrical grid must be robust enough to withstand disruptions and adaptable enough to incorporate new technologies. Enhancing grid resilience and reliability involves a multifaceted approach that includes infrastructure upgrades, advanced technologies, and strategic planning.

The traditional electrical grid, designed for a predictable and centralized flow of electricity, is being challenged by the growing presence of distributed energy resources (DERs) such as solar panels and wind turbines. These renewable sources, while environmentally beneficial, introduce variability and intermittency into the grid. To address these challenges, grid operators are turning to advanced technologies that enable real-time monitoring and management of the grid. By leveraging data analytics and predictive modeling, operators can anticipate fluctuations in energy supply and demand, optimizing the use of renewable energy and maintaining grid stability.

One of the key strategies for enhancing grid resilience is the implementation of microgrids. These localized energy systems can operate independently or in conjunction with the main grid, providing a reliable power supply during outages or disruptions. Microgrids are particularly valuable in areas prone to natural disasters, where they can ensure a continuous power supply to critical infrastructure such as hospitals and emergency services. By incorporating renewable energy sources and energy storage systems, microgrids contribute to a more sustainable and resilient energy system.

Energy storage is another critical component of grid resilience. By storing excess energy generated during periods of low demand, storage systems can release energy during peak demand, smoothing out fluctuations and ensuring a stable power supply. Advances in battery technology, such as lithium-ion and flow batteries, are making energy storage more efficient and cost-effective. These systems not only enhance grid reliability but also facilitate the integration of renewable energy sources, reducing reliance on fossil fuels and lowering carbon emissions.

The modernization of grid infrastructure is essential for improving resilience and reliability. Aging infrastructure is more susceptible to failures and outages, posing a risk to the stability of the grid. Upgrading transmission and distribution lines, substations, and transformers can enhance the grid's ability to withstand physical and cyber threats. Additionally, the deployment of smart grid technologies, such as advanced metering infrastructure and automated control systems, enables real-time monitoring and rapid response to grid disturbances, minimizing the impact of outages and improving overall reliability.

Cybersecurity is a critical consideration in the effort to enhance grid resilience. As the grid becomes more interconnected and reliant on digital communication, it becomes vulnerable to cyber threats. Protecting the grid from cyberattacks is essential to ensure the security and reliability of electricity services. Implementing robust cybersecurity measures, such as encryption, authentication, and intrusion detection systems, is crucial for safeguarding the grid against potential threats. Ongoing research and collaboration among industry stakeholders are essential to staying ahead of evolving cyber threats and ensuring the continued security of the grid.

The role of policy and regulation in enhancing grid resilience and reliability cannot be overstated. Governments and regulatory bodies play a crucial role in setting standards and providing incentives for infrastructure upgrades and the adoption of advanced technologies. Supportive policies, such as tax incentives and grants, can encourage investment in grid modernization and the development of resilient energy systems. Additionally, regulatory frameworks that promote competition and innovation can drive the development of new technologies and solutions that enhance grid resilience.

Collaboration among utilities, technology providers, and research institutions is essential for advancing grid resilience and reliability. Public-private partnerships can provide the necessary funding, expertise, and infrastructure to implement innovative solutions and bring them to scale. By working together, stakeholders can overcome the challenges associated with grid modernization and unlock the full potential of advanced technologies.

The integration of renewable energy sources into the grid presents both challenges and opportunities for enhancing resilience and reliability. While the variability of renewables can pose challenges for grid stability, their integration can also enhance resilience by diversifying the energy supply and reducing reliance on centralized power plants. By optimizing the use of renewable energy and incorporating energy storage systems, grid operators can create a more flexible and resilient energy system that can adapt to changing conditions and withstand disruptions.

The future of grid resilience and reliability is promising, with ongoing research and development driving continuous improvements in technology and infrastructure. As technology continues to evolve, new and innovative solutions are expected to emerge, further enhancing the capabilities and benefits of the grid. The transition to a more resilient and reliable grid is a complex and multifaceted endeavor, requiring collaboration among utilities, regulators, technology providers, and consumers. By working together, stakeholders can overcome the challenges associated with grid modernization and unlock the full potential of advanced technologies.

The journey towards a more resilient and reliable grid is marked by challenges and opportunities, but with continued commitment and collaboration, the potential to transform our energy systems is immense. By harnessing the power of advanced technologies and strategic planning, we can create a more efficient, reliable, and sustainable energy system that meets the needs of future generations.